Water

A portfolio of thirty original watercolours exploring the sensory and aesthetic properties of water through the lens of science and ancient philosophy, expounding the art of being curious

ROBIN MOORE

"If there is magic
on this planet,

"...if there is magic on this planet,
it is contained in water."

Loren Eisely, anthropologist

The kaleidoscope of colours created at sunset on the beach at Polzeath, Cornwall

AuthorHouse™ UK
1663 Liberty Drive
Bloomington, IN 47403 USA
www.authorhouse.co.uk
UK TFN: 0800 0148641 (Toll Free inside the UK)
UK Local: 02036 956322 (+44 20 3695 6322 from outside the UK)

Because of the dynamic nature of the Internet, any web addresses or links contained in this book may have changed since publication and may no longer be valid. The views expressed in this work are solely those of the author and do not necessarily reflect the views of the publisher, and the publisher hereby disclaims any responsibility for them.

This book is printed on acid-free paper.

ISBN: 979-8-8230-8034-7 (sc)
ISBN: 979-8-8230-8033-0 (e)

Library of Congress Control Number: 2023908159

Print information available on the last page.

Published by AuthorHouse 05/01/2023

authorHOUSE®

WATER: a transparent, odourless, tasteless liquid, freezing at 0° C and boiling at 100° C. It contains 11.188 per cent hydrogen and 88.812 per cent oxygen by weight.

Earth's approximate water volume, the total water supply of the world, is 1.386×10^9 cubic kilometres (3.33×10^8 cubic miles).

Without it, we would not be here!

 The ancient Greek symbol for water

Preface

Curiosity is the state of being curious: inquisitive, wondering, ready to poke around and figure something out. Which, apparently, I am.

The word used to mean "very, very careful", which I am not!

Only in the last few hundred years has it changed into a word expressing the desire to know more. Which I do.

My curiosity about water springs from my first book, *The Biography of the Village of Little Houghton*. My research into why the village started where it did and still exists today sent me on a journey to the last ice age, circa 13500 BCE, where I discovered ice a mile thick changing the shape of the land on which the village of Little Houghton grew. The last ice age and its subsequent retreat created a great lake of meltwater, which flooded and formed a river. Over time, a crossing point on the river became a major north-south highway in England. A village was created at the crossing point called Little Houghton, or the village on a spur. Without the effects of the ice age, the village of Little Houghton might never have existed.

I wrote my second book about my great-grandfather's journey in 1887, when he and his family sailed from London to Buenos Aires by steamship to take part in building the docks that would enable Argentina to trade with other countries across the world's oceans. This story again highlighted the importance of water and the impact it has on our lives.

It started me thinking. What is water? Where did it come from? When did it cover the earth? Why is it still here? How does water create our world's climate? And not to forget the obvious question, why is it blue? Naturally, we see and use water every day. It is all around us. However, little did I appreciate how miraculous and special this chemical compound known as H_2O is—until I started poking around to create this portfolio of watercolours.

This portfolio is not about the science of water, because this simple compound is too vast and elusive to be contained in a single volume. Nor is it a textbook on the study of water. The subject is too wide and interdisciplinary, covering many "ologies". Learning about water is a lifetime's endeavour, and fortunately for me, a lot of very clever academics and scientists have done that already. However, in my quest to create a portfolio illustrating the way water has a profound impact on our emotions, I too found that I had to understand the natural laws of water before I could illustrate it successfully in the medium of watercolour.

My investigation into the way water flows and moves, creates patterns, and shapes our lives took me on an interesting journey from the humble drip to the mighty oceans. I have tried to explain the history and science that helped me create my portfolio and to highlight some interesting facts about water that I discovered on my journey.

Most of all, my portfolio of thirty watercolours is meant to highlight the sensory and aesthetic impact water has on our well-being. Art is about emotion, understanding colour, and observation. I hope that bringing art together with a little science will help us understand water for the power that it is—and appreciate its true magic.

I hope they make you curious too.

Introduction

My journey to make sense of the emotional properties of water is loosely based on the conclusions of the early philosophers that everything on earth was made up of four things: water, air, earth, and fire.

The early Greek philosophers were searching for the first principle in life, or roots—in our terms, the smallest part that made up everything in life.

Aristotle was the first to call such a root *stoicheion* (στοιχεῖον), the smallest unit of time on a sundial, or an indivisible unit. Democritus and Epicurus called this unit an *atomon*, meaning that which cannot be cut or divided, hence the word "atom"—a word we still use today.

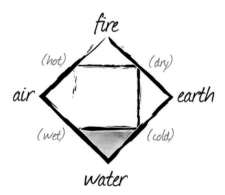

Figure 1   The relation of the four elements

The ancient Greek concept of four basic elements—earth, water, air, and fire—dates from pre-Socratic times and persisted throughout the Middle Ages and into the Renaissance, deeply influencing European thought and culture.

The importance of water is supported by the hierarchical charts of the time (Figure 1), with water supporting earth, air on top of earth, and fire or the sun over everything.

Empedocles was one of the first Greek philosophers to believe that everything was made up of the four elements of water, earth, air, and fire. This theory of the importance of water was suggested around 450 BCE and was later supported and added to by Aristotle.

The Greek philosopher Thales suggested that water was the ultimate underlying substance from which everything is derived; Anaximenes subsequently made a similar claim about air. However, none before Empedocles proposed that matter could ultimately be composed of all four elements in different combinations.

Other great philosophers and scientists of their time have also tried to make sense of our world.

The Persian philosopher Zarathustra (600–583 BCE) described the four elements of water, earth, air, and fire as "sacred", saying that they were "essential for the survival of all living beings and therefore should be venerated and kept free from any contamination".

In Babylonian mythology, the cosmogony called _Enûma Eliš_, a text written between the eighteenth and sixteenth centuries BCE, involves four gods that we might see as personified cosmic elements: Sea, Earth, Sky, Wind.

Five elements are found in the Hindu system of the Vedas, especially the Ayurveda. The _pancha mahabhuta_ or "five great elements" of Hinduism are water, earth, fire, and air divided into wind and aether.

In Buddhist Pali literature, the _mahabhuta_ or "great elements" are earth, water, fire, and air. In early Buddhism, these four elements are a basis for understanding suffering and for liberating oneself from suffering. I have used this way-of-life philosophy as a theme in my watercolours to bring out the revolving and constant state of life-giving water.

The Chinese had a somewhat different series of elements—namely, water, wood, fire, earth, and metal (literally gold)—understood as different types of energy rather than, as in the Western tradition, kinds of material. These energies are in a state of constant interaction and flux with one another.

It's amazing that the idea of water and the other elements making up all matter was the cornerstone of philosophy, science, and medicine for two thousand years. From the perspective of modern science, of course, these theories are inadequate and misleading. But in a way, the four elements do align with the four states of matter that modern science has agreed on: *liquid* (water), *solid* (earth), *gas* (air), and *plasma* (fire).

For me, these ancient theories hold poetic resonance, which is a different kind of truth from the hard facts of science. We may respond emotionally or spiritually to that truth while still being amazed by our scientific progress represented by the elements on the periodic table.

My hope is that my watercolours may enhance the emotional, sensory, or aesthetic experience of water. Something that makes us stop for a moment in our busy lives and consider the emotional impact of water, something that we take for granted most of the time. Enjoy!

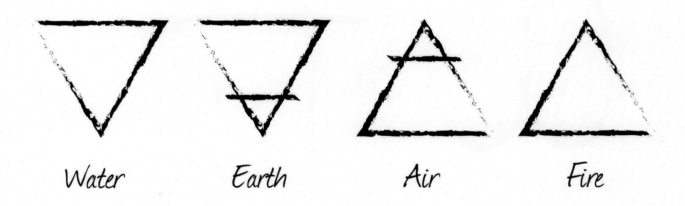

Figure 2   Empedoclean symbols for the elements

Water

Reflections on a lake the three states of water as liquid, solid, and vapour

PART 1

The Magic of Water

The start of our magic show

The inspiration of ever-changing cloud formation

Water has been a subject of fascination for a very long time.

From the seventh century BCE, when Thales suggested that water was the ultimate underlying substance from which everything is derived, many philosophers have studied the emotional connection we have with water.

Leonardo da Vinci was fascinated by water and carefully observed its "stickiness".

In the early fifteenth century, Arab navigator Ibn Majid compiled his extraordinary work *The Book of Profitable Things Concerning the First Principles and Rules of Navigation*. In it, he recorded his observations of the movement of water, which he used to find his way around the oceans long before navigational maps were drawn up.

Early scientists and pioneers became experts in navigation and predicting weather from the movement of water and the effect weather has on our lives.

Today, many research projects are underway to discover why water has a powerful physiological effect on our well-being and emotional health.

A large cross-sectional study of 3,327 people, published in the *World Journal of Psychiatry*, concluded that drinking water is associated with decreased risk of depression and anxiety in adults.

The study of water is an enthralling, never-ending show.

Just a drop in the ocean?

Hydrology is the study of the movement, distribution, and quality of water throughout Earth.

The collective mass of water found on, under, and over the surface of a planet is called the hydrosphere.

Movement of water throughout the hydrosphere is responsible for the distribution of both heat energy and nutrients, regulating the temperature and making earth habitable for us. Water never stops—it could be called the world's Magic Circle.

Perpetual Motion

Wherever we find water anywhere across the globe, its nature is constantly changing.

In a hundred-year period, the average water molecule spends ninety-eight years in the ocean, twenty months as ice, about two weeks in lakes and rivers, and less than a week in the atmosphere before falling as rain and starting the cycle all over again.

Throughout the ninety-eight years in the ocean, thermal currents, tides, waves, and wind keep the water molecules in perpetual motion.

The water cycle is responsible for the distribution of heat and nutrients, regulating the temperature on earth and shaping the land.

The continuous exchange of water within the hydrosphere—the atmosphere, soil water, surface water, and groundwater—supports all life. Water moves perpetually through the stages by means of these transfer processes:

- **evaporation** from oceans and other water bodies into the air
- **transpiration** from land, plants, and animals into the air
- **condensation** from water vapour in the air to liquid water, occurring as evaporated water rises into the atmosphere and cools
- **precipitation** as condensing water vapour in the air falls to the ground or ocean as rain, sleet, snow, or hail
- **run-off** over the land as water collects in a large water body, eventually reaching the ocean
- **infiltration** as precipitation that does not run off moves deep into the soil, seeping down into underground water stores
- **solidification** as liquid water freezes into solid form, creating ice and glaciers

The challenge of capturing ever-moving water is to never stop looking.

The magic circle sea, evaporation, cloud, and rain

19

Lakes create their own life systems and climate.

EVAPORATION

When precipitation collects in an enclosed area to form a lake, the body of water can create *katabatic wind*—moving air that travels down the side of the surrounding hills, carrying away water vapour. "Dry" air takes its place and moves over the surface water to increase evaporation.

Evaporation from open water, such as lake surfaces, is a critical and continuous process in the water cycle.

Sometimes you need to climb to a high vantage point to fully appreciate the size, shape, and grandeur of a lake. The evaporation from surface water can be identified by the subtle blues and violets in the haze of distant objects.

A broadleaf tree can transpire over 230 litres of water per hour into the air on a hot summer's day.

TRANSPIRATION

One acre of broad-leaved forestry has the ability to release up to 36,000 litres of water into the atmosphere every summer's day. Incredible!

When you consider that one cubic metre (one thousand litres) of water weighs one tonne, you can appreciate what a huge weight is carried up through the tree from the roots to the top leaf on the crown, often more than thirty metres from the ground. To move water from the roots to the crown, a continuous column of water must form—an idea known as *cohesion theory*.

The polarity of water molecules both attracts and repels nearby water molecules to maintain this continuous column.

This cohesive column of water is thought to be initiated when the tree is a newly germinated seedling. It is maintained throughout the lifespan of the tree by two forces: *osmosis* or *capillary action*, and *evapotranspiration*.

- Osmosis or capillary action draws water into the roots from the soil, and forces water to rise up the plant through root pressure.
- Evapotranspiration draws water up to the crown, keeping water moving up through the tree as it evaporates from leaves at the top.

A majestic oak in Little Houghton, dappled light and dark from the changing light on a peaceful summer's day, disguises the fact that this tree is silently working very hard.

New leaves in spring: more than just a breath of fresh air

The magic factories of life—
releasing oxygen for us to breathe.

PHOTOSYNTHESIS AND CELLULAR RESPIRATION

Water is fundamental to photosynthesis in plant life. Photosynthetic cells use the sun's energy to split off hydrogen from the water molecule. The hydrogen is used to form glucose molecules in the plant by combining with carbon dioxide, which is absorbed from air or water. The remaining oxygen is released into the atmosphere or used by the plant in cellular respiration.

Mist, locally known as haar, over the Outer Hebrides

Layers of cold air and warm, moist air create mist, fine droplets of condensed water that hang in the air.

CONDENSATION

Mist that forms on a still, windless day acts as a light filter. The fine water droplets diffuse the light from the sea and sky, creating a soft glow, as for the haar seascape in the Outer Hebrides.

It takes millions of water molecules to form one raindrop.

PRECIPITATION

Watching rain falling in the distance, you notice how it affects the transmission of light through the sky-brightness and colour change according to the intensity of rainfall.

With wind gusting the clouds and the displacement of the air by the falling water, wonderful skyscapes are formed.

Rain clouds creating an ever-changing pattern.

Water journeys back to the ocean in our rivers.

RUN-OFF AND INFILTRATION

Precipitation that does not run off infiltrates deep into the soil, seeping down into underground water stores. Rivers are formed when the precipitation rate exceeds the saturated conductivity of the soil and aquifers are full.

It appears that no matter how big or small a natural river is, they all have the same set of features in the same sequence like this river in The Scottish Borders.

If you walk a hundred metres along a natural river that is ten metres wide, you are likely to pass two riffle, glide, or pool sequences.

READING RIVERS: THALWEGS, EDDIES, RIFFLES, POOLS, GLIDES, AND OTHER DELIGHTS

Looking at a natural river, you soon realize that the water will not run straight for long—not more than ten times its own width, in fact! Rivers continually meander, twist, and turn as water changes its course in response to obstructions along the riverbanks and in the riverbeds. Over time, various names have been given to these features.

A thalweg is the line in the river where the water flows fastest, which leads to several other features.

Eddies form where fast water passes something fixed like a riverbank, where it slows, spins, and causes some part of the water to flow upstream.

Riffles are easy to identify because they change the vertical height of the water. An obstacle in shallow water, like a rock on the riverbed, forces water to move up and over the top, creating foaming whiter water. This creates a bulge of water called a pillow before the rock, and a hole after it called a stopper, where the water dips, flows upstream, and splashes.

Pools are just as easy to spot, as they are deep and typically calm with a smoother surface. They are found where the water has gouged out the riverbed.

Glides are found between the riffle and pool where the water flows quickly and smoothly.

The light reflecting off frozen water provides a colourful, glassy sheen, an effect that does not occur with water in its liquid state.

SOLIDIFICATION

Water is unique, as it is the only compound that is less dense as a solid than a liquid.

As water turns solid, its volume increases by 9 per cent. This expansion can exert enormous pressure, bursting pipes and cracking rocks.

Water differs from most liquids in that it becomes less dense as it freezes. When water solidifies at atmospheric pressure, it reaches its maximum density of 1,000 kg/m3 at 3.98°C. In a lake or ocean, water at 4°C is at its densest, so it sinks to the bottom, and cooler water forms ice with a lower density, 917 kg/m3, which sits on the surface.

The fact that ice floats on water is exceedingly important in the natural world, because the ice that forms on ponds and lakes in cold areas acts as an insulating barrier to protect the aquatic life below. If ice were denser than liquid water, the ice forming on a pond would sink, thereby exposing more water to the cold air temperature. The pond would eventually freeze throughout, killing all the life forms present.

The sea often freezes in the Arctic Circle, as at Tromso, Norway, but as the ice floats on the surface, the sea life remains in the liquid water below, and fishing will continue in the summer.

Fishing sheds on a frozen fjord to the north of Tromso, Norway;

the light reflecting off the frozen water produces a colourful, glassy
sheen and a reflective quality not seen on liquid water.

Our island has had at least eight ice ages. The last started retreating about ten thousand years ago.

ICE AGES

We are very lucky that human civilization could develop during an era of Earth's low orbital eccentricity.

In a rhythm attuned to regular wobbles in Earth's orbit and spin, ten eras of spreading ice sheets and falling seas have come and gone over the last million years, eight affecting Great Britain. That seems to indicate an Earth wobble every hundred thousand years, causing an ice age separated by interglacial periods of between ten and thirty-five thousand years.

Cold spells have so dominated Earth that geophysicists regard warm periods like the present one, called the Holocene, as the oddities. The scientific name for these periods—interglacial—reflects the exceptional nature of such times.

It appears that scientists think the next ice age will almost certainly reach its peak in about 80,000 years, but debate persists about how soon it will begin. The latest theory is that human influence on the atmosphere may substantially delay the transition. Otherwise, we might already be heading for the next ice age!

I find it hard to visualize ice a mile thick over England during the last ice age and terminal lakes the size of Norfolk!

ICE AND GLACIERS

In the last ice age, two thirds of Britain and Ireland were covered by a thick ice sheet extending from the west coast of Ireland and across the North Sea all the way to Norway. The British ice sheet reached as far south as Oxfordshire and Essex, diverting the original course of the River Thames, which had flowed north to the Wash, to its current position through what is now London.

When glaciers began their final retreat from England about ten thousand years ago, they left behind many landscape features, including lakes, valleys, and mountains as well as rocks from Norway. You can often see the direction of flow of a glacier by observing clues left behind in the landscape.

Glaciers are rarely smooth and create a particular challenge to a painter. The pure brilliance of ice absorbs, reflects, and refracts light in all directions, creating unusual patterns of light and shade.

PART 2

Surprising Facts about the Life of Water

It is worth taking a few moments in our busy lives to stop and enjoy the world of nature around us. The sheer scale and complexity of water is astounding.

However, what stirs the emotions within us is the magic of water.

Roughly 70 per cent of the Earth's surface is covered in liquid water.

The Earth's surface is made up of 66 per cent ocean and 29 per cent land, with inland lakes and rivers making up 5 per cent.

I find it remarkable, considering the size of planet Earth, that the surface is covered with 70 per cent water and roughly 30 per cent of surface land, a proportion which has remained relatively constant over billions of years.

It is always a challenge to capture the image of movement in water: the huge horizons, the scale from raging oceans to reflective dead calm, from the smallest droplet to towering clouds and the gentlest of breezes.

My challenge in this watercolour was to capture the chaos caused by the ocean swell hitting the sandbar in the mouth of the Camel Estuary in Cornwall. Many a ship has sunk on this sandbar, driven by the turbulent chaotic currents.

This watercolour is about the contrast between the order of the ocean shattered on the bar and the instant change of colour and luminance caused by the coming storm.

50

Our atmosphere has 67 per cent cloud cover on average— nearly equivalent in area to the ocean's surface.

Water evaporates from the ocean surface and transpires from plants. Water vapour rising from the surface eventually condenses in the upper atmosphere to form clouds.

It's a striking fact that the volume of global average cloud cover is constant across the Earth's surface and nearly in the same proportion as the water cover.

Low-level clouds have a net cooling effect on the Earth's surface because they reflect solar energy. They trap infrared radiation at night, reducing temperature fluctuation, but the net effect is cooling. This process maintains our unique climate and our water supply.

The big skies that occur over the Yorkshire Moors are an ideal place to capture the subtle colours in clouds.

54

55

Over 60 per cent of the world's supply of fresh water is stored in Antarctica in the form of ice.

Did you know that the 7 million square miles of Antarctica hold more fresh water than any other place on Earth? The Antarctic ice cap contains about 90 per cent of all the ice in the world, and about 80 per cent of all fresh water in the world is in the form of ice. The area of sea ice around Antarctica extends in the winter, doubling the continent's size.

Sea ice forming around Antarctica

The water that waves leave behind, known as shallow wash, creates a kaleidoscope of reflective colours that constantly change–your very own light show, best seen when looking towards the sun on a cloudy day, as on this beach in Cornwall.

When I stood on a beach, watching the waves roll in and splash about, I was humbled to think that I was looking at water that may have been created 4.6 billion years ago.

When did water cover Earth? As far as scientists can say, liquid water did not exist on the planet when it was first formed.

Scientists suspect that our planet formed dry, with high-energy impacts creating a molten surface on the infant Earth. Water came much later as Earth cooled, about 4.6 billion years ago.

Surprisingly, the water that I was looking at may have originally arrived from the outer limits of our universe.

Scientist theorize that asteroids are most likely the primary source of Earth's water. Carbonaceous chondrites, a subclass of the oldest meteorites in the solar system, have hydrogen isotopic levels most similar to Earth's ocean water. Two 4.5-billion-year-old meteorites found on Earth, formed 14 million years after the birth of the universe, contained liquid water alongside a wide diversity of deuterium-poor organic compounds, further supporting this theory.

Researchers estimate that the equivalent of as many as ten oceans of water may exist within the Earth's mantle. When the mantle's heat causes hydrogen- and oxygen-rich minerals to melt, the resulting water spews from the planet's crust. This water, in the form of vapour and other gases, escaped from the molten rocks of the Earth as it was still forming, into the atmosphere surrounding the cooling planet. After the Earth's surface had cooled to a temperature below the boiling point of water, rain began to fall—and continued to fall for centuries.

Additional evidence from the University of Münster in 2019 shows that the molybdenum isotopic composition of the Earth's core matches that of the outer solar system. Their explanation is that Theia, the planet said to have collided with Earth 4.5 billion years ago, resulting in the formation of Earth's moon, may have originated in the outer rather than the inner solar system, bringing water and carbon-based materials with it.

Looking west, watching the sun rise over the Indian Ocean.

Sunrises, like this one in Sri Lanka, create a soft light owing to the mists that form over water before the air is warmed by the sun. The mists create a diffusing atmosphere through which the sun's rays must pass before being caught by your eye.

Water from the Earth's mantle is released into the atmosphere via volcanoes.

Earth is the only planet known today to have oceans of liquid water on its surface. Liquid water, which is necessary for life as we know it, continues to exist on the surface of Earth because the planet is at a distance from our sun known as the habital zone—far enough away that it does not lose its water into space, but not so far that low temperatures cause all water on the planet to freeze.

It is thought that additional water, approximately three times the mass of Earth's oceans, could be stored in the mantle. Similarly, Earth's core could contain four to five oceans' worth of hydrogen.

Today the atmosphere is rich in oxygen, which reacts with both hydrogen and deuterium released from Earth's mantle via volcanos to recreate water, which falls back to the surface.

Volcanoes form at subduction zones along the coastline where tectonic plates are moving towards each other and one plate descends beneath the other.

Subduction provides a mechanism for introducing water-bearing sediments into the mantle. As the subducted oceanic plate sinks and heats up, water is released gradually from the sediments and minerals within the plate slab. Water has the effect of reducing the melting temperature of the mantle by about 60° to 100°C. It is this process that allows the generation of magma at a depth that feeds volcanoes, which are formed at the surface. Volcanoes deliver water from the mantle into Earth's atmosphere.

Therefore, the vast bulk of water on Earth is held in a closed system that prevents the planet from gradually drying out. What little water is lost into space is replaced from Earth's mantle.

There are more than two thousand volcanoes in Chile alone. Five hundred are potentially active. Sixty have erupted over the last 450 years. This one in Villarrica, Pucon, is one of the most active.

Waves can travel thousands of miles, yet the water in them hardly moves.

A sizable wind-created wave can travel thousands of miles, but the water molecules in the wave hardly move forward at all. Why?

The water molecules in the wave orbit in a circle, while the energy in the water travels horizontally forward in the direction of the wind. The water molecules under the surface circle in turn orbit in a slightly smaller circle, and the water under that also orbits in a slightly smaller circle, until there is no wave movement in the water.

The visual effect on the surface is the crest and trough of a wave moving forward, but all the water molecules are doing is rising and falling in circular motion.

I find it interesting that winds are created by our sun heating Earth and its atmosphere from 92 million miles away. Those winds transfer their energy to water and create waves. The waves transport the energy to a distant shore, shaping the land as the wave breaks.

Water is continually moving sand, silt, and pebbles along our coastline every day, changing and reshaping our land.

Pebbles, shingle, and sands move along our coastline in a very predictable way. They never stay still for long. Longshore drift is the action of waves rushing up the beach at an angle and then retreating directly down the beach by gravity, assisted by the ebb and flow of tides. This action moves the loose pebbles, shingle, and sand many miles along the coastline until they reach a natural barrier.

Around our island, generally the water moves pebbles, sand, and silt from north to south along the East Coast, east to west along the South Coast, and south to north along the West Coast—which means a pebble formed on the western coast of Cornwall could be found much further north. Local winds and tides may cause a few local variants in direction.

Sandbars and dunes on the Camel estuary and bay change with every tide.

PART 3

An Investigation into the Colour of Water

And why is the sky blue?

Why are clouds white?

Why are sunsets yellow, orange, or red?

Why is the sea blue, green, or grey?

Why does clear water act like a mirror?

Glasto

It's fair to say that water has little colour of its own. When you look at a glass of water, it appears both colourless and clear.

The variety of colours we see when we look at the sea, sky, rivers, or lakes makes it hard to describe the colour of water in its natural state. Any colours seen are more likely the result of the reflective quality of water or of what is under or in the water itself. Ultimately, the colour perceived comes down to how water reflects the sun's light, which creates the vast range of colours we see.

The early Britons used the colour word glasto to describe water, denoting things that were in the blue-green-grey range. It seems like a good description to me!

Interestingly, the modern English word glass is probably related etymologically to glasto.

When you look at water from overhead there are no reflections. You can then clearly see the colours of the water change according to the depth of water, as in this aerial view of a tropical beach. Small particles that waves carry will also change the colour of the water.

And why is the sky blue?

Thanks to the explanations of Michael Kruger of the Department of Physics, University of Missouri, I learned during my investigations that the sky is blue because the atmosphere tends to scatter shorter-wavelength blue light more than longer-wavelength red light.

Since blue wavelengths of light from our sun are scattered much more than the other colours, you see blue when you look up at the daytime sky no matter where you look.

This effect called *Rayleigh Scattering* after its first theorist, Lord Rayleigh, is most prevalent when the particles that do the scattering are smaller than the wavelength of light, as is the case for the nitrogen and oxygen molecules in the atmosphere.

Why are clouds white?

The scattering of light from the sun also makes clouds look white. Clouds are white because light from the sun is white, but in a cloud, sunlight is scattered by much larger water droplets. These larger droplets scatter all colours almost equally, so that the sunlight remains white and makes the clouds also appear white against the background of the blue sky.

When clouds are thin, they let more light through, therefore appearing white. But like any objects that transmit light, the thicker the clouds are, the less light makes it through. As their thickness increases, the bottoms of clouds look darker but still scatter all colours. We perceive this as grey. If you look carefully, you will notice that the relatively flat bottoms of clouds are always a little greyer than their sides. The taller the clouds become, the greyer their bottoms look.

White clouds in the sky and their reflection in the water of a river

Why are sunsets yellow, orange, or red?

When the sun is low on the horizon, sunlight passes through more air than during the day, when the sun is higher in the sky. More atmosphere means more molecules to scatter the violet and blue light away from your eyes. This is why the sky is often yellow, orange, and red at sunset and sunrise.

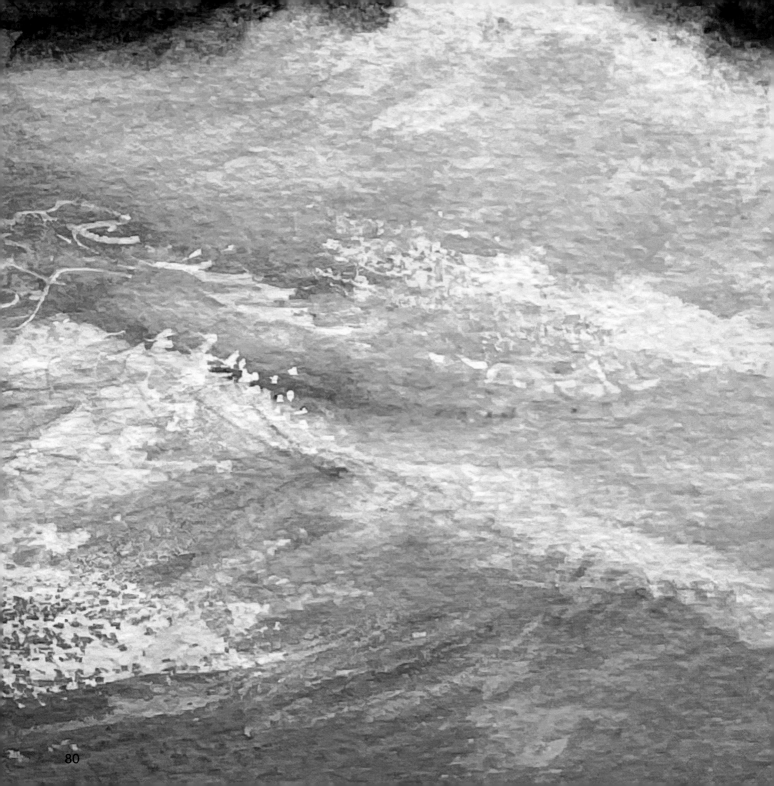

Why is the sea blue, green, or grey?

The colour of seawater is determined by a different process, as I learned from the work of Jennifer Levine, a specialist with the National Ocean Service. The ocean is blue because water molecules absorb light, like a filter, in the long wavelengths of the visible spectrum, leaving behind light waves in the short-wavelength blue part of the light spectrum. The blue light is then reflected for us to see. The darker the blue, the deeper the water.

The ocean may also take on green, red, or other hues as light bounces off floating sediments and particles in the water as well as off the seabed in shallow water.

Most of the ocean, however, is completely dark. Hardly any light penetrates deeper than 656 feet, and no light penetrates deeper than 3,280 feet, so water will look black.

Why does clear water act like a mirror?

Water is reflective mainly because it is relatively flat, and its index of refraction is different from air. However, the ability of water to reflect light varies depending on the angle at which light strikes the surface of the water, commonly known as the angle of incidence, and the height of the eye above the water, the angle of reflection.

Seawater reflects only 2 or 3 per cent of light when the angle is less than 50 degrees. That's about the same as glass, so you can see through the water. At greater than 50 degrees, the reflective cadence increases, so that light reflects off the surface of the water and you can see reflections of the surrounding area on the water.

Interestingly, a fish underwater would experience a similar phenomenon. Outside its sky circle—the area a fish can see through the water to whatever lies beyond—a fish sees a reflection of the seabed and objects in the underside of the water's surface.

This image of a lake shows how the reflection of trees on the far bank can be seen in the water when the viewer is standing in the correct position opposite.

I have described water in the oceans, in
the sky, and in rivers and lakes, but a
simple drop of water has colour too—and
the science it contains is amazing.

Curious ...

Have you noticed that a droplet of water on a flat surface forms a "bubble" all on its own?

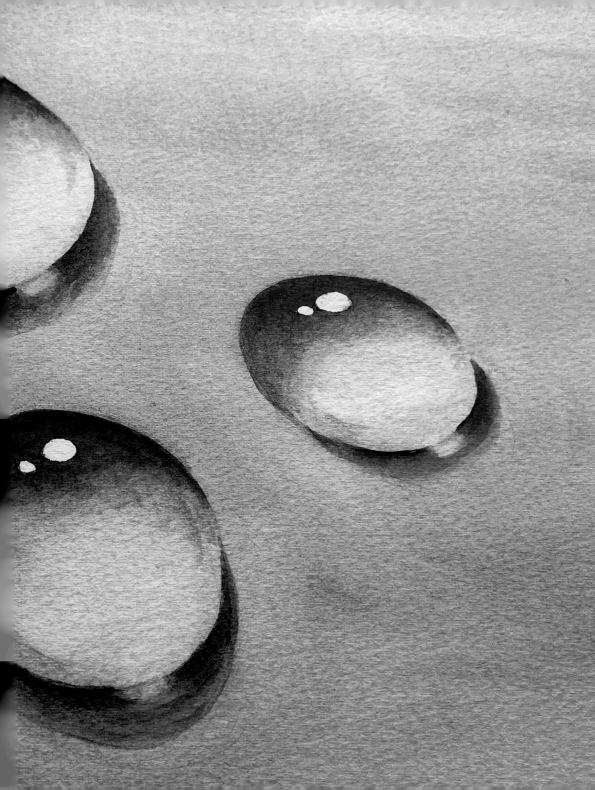

JUST A HUMBLE DROPLET OF WATER

Why does a droplet of water form a "bubble"?

Water molecules are made up of two hydrogen atoms and one oxygen atom. The covalent bonds that form between hydrogen and oxygen atoms, which differ in size, result in a molecule that has a net positive charge near the hydrogen atoms and a net negative charge near the oxygen atom.

This difference in charge within a water molecule allows the oppositely charged parts of a water molecule to make weak hydrogen bonds with the neighbouring water molecules which hold a body of water together.

The cohesion of water molecules helps plants take water up through the xylem (conductive tissue) in the stem of plants. This cohesion also contributes to water's high boiling point and surface tension.

It is surface tension that stops water from spreading out, and it is also the reason water forms drips from a tap and raindrops.

Surface tension is also why colours form inside a droplet and why light becomes trapped inside.

Next time you see a droplet on a surface, take a close look and notice the way colours and light make it shine.

Notice that when light enters a droplet, it is reflected off the inside of the edge opposite where it entered, causing an internal highlight.

Light entering a droplet also creates an internal refracted shadow on the inside surface opposite. This highlight then casts a shadow on the outside of the drop onto the surface it is lying on. The ray of light that entered the drop exits directly opposite where it entered, creating a light on the surface within the shadow.

Oh, and notice the highlight where the light entered the drop in the first place. This is the reverse of what you would expect from light falling on a solid object or on a clear one like glass.

When I set out on my journey of discovery, I found that water has a set of rules. The way light interacts with water is different to the way light falls on other objects, and that makes it difficult to paint. That effort to understand why was my introduction to the extraordinary properties of water, and my inspiration to create this portfolio of watercolours.

Having explored some of the properties of water, is it right that a humble drop is taken for granted, when actually it is a polar molecule with extraordinary properties that is essential for all life on Earth?

Acknowledgements

There is a lot more to water than I expected when I first had the idea to paint it. That idea sent me on a voyage of discovery, through many stages of "Why does it do that?" My understanding of the science of water was pretty sketchy—a lot of "Wow! Really?"

And so, I am very indebted to Eleanor, my daughter, who not only understands the science way beyond my simple descriptions in this portfolio but has the interest and communication skills to take me through the tricky bits. Her promptings—"Are you sure?" and "Is that right?"—encouraged me to double-check and simplify some of my explanations.

A big to thank-you to Cynthia Wolfe, my editor at AuthorHouse who provided insight into the art of making a book flow, among other things. Thank-you also to the people helping behind the scenes, who showed remarkable patience laboriously correcting my grammar and punctuation of the text and generally putting up with me through change after change, carefully crafting the artwork as I tried to make the story flow, as well as to my grandchildren, with whom I had endless fun playing with paint and experimenting. They showed no fear of doing things differently!

There are many others, great scientists and academics, who have helped and guided my research without knowing it. Many have spent their life studying water, publishing their findings, and making that knowledge available to all through the World Wide Web. Without their unknowing help, I would not have been able to produce this book. Thank you all.

I must not forget all those who have given me help and encouragement to keep going with this project—those whom I have bored silly with my "Did you know?" exclamations when I found out some fact or bit of science that I found extraordinary.

I know I will have made mistakes, and they are all mine.

Printed in the United States
by Baker & Taylor Publisher Services